# The Art of Designing Organic Reaction Mechanisms

### Editor

Dr. Sarosh Iqbal

**SHAMA BOOKS**
PUBLISHERS & BOOK SELLERS
• PAKISTAN • INDIA • ENGLAND

Since 1965
GOLDEN JUBILEE

I0048340

Copyright © SHAMA BOOKS

| | |
|---|---|
| Book Name: | The Art of Designing Organic Reaction Mechanisms |
| Editor & Authors: | Dr. Sarosh Iqbal |
| | Dr. Tahsin Gulzar |
| | Dr. Tahir Farooq |
| Distributor: | Muhammad Ajmal Hameed |
| Edition: | 1st-October, 2017 |
| ISBN: | 978-969-658-025-6 |
| Price: | 350-PKR |
| | 30-US$ |

Publishers

# SHAMA BOOKS

Tel: +92-41 2627568, 2613449, +92-302 7175174
www.facebook.com/ShamaBooks

© All rights reserved. No part of this publication may be restored, stored in retrieved system or transmitted in any forms by means of electronics photocopying, recording etc. without written permission of the author (s) and publisher (Copyright holders).

# Preface

There are numerous organic chemistry books available that covers material from basic to advance level. Most texts are factually based but students are unable to apply this factual knowledge to solution of problems related to organic chemistry. This book has been designed to make students capable enough that they can apply their basic knowledge of organic chemistry to solve the problems. In this book, authors have tried to demonstrate that there are set of rules which can be grasped with a little effort and without any doubt, use of these rules in a systematic manner reduces the burden on memory! ☺

This book has been categorized into following chapters

Nucleophilic Substitution Reactions

Nucleophilic addition to carbonyl groups

Elimination Reactions

Aromatic Compounds

All chapters cover brief introduction, a complete step-wise strategy to solve the problems and practice problems with hints. Answers of all practice problems are given in the end of book. Students are supposed to solve practice problems by themselves using hints and step-wise strategy, in the given space.

**Sarosh Iqbal**
(Ph.D.)

*This humble effort is dedicated to my beloved Parents*

*Mr. Muhammad Iqbal*
*&*
*Mrs. Nasim Iqbal*

# Table of Contents

|  |  | Page No. |
|---|---|---|
| **Chapter-1** | Nucleophilic Substitution Reactions | **01** |
| **Chapter-2** | Nucleophilic addition to carbonyl groups | **17** |
| **Chapter-3** | Elimination Reactions | **40** |
| **Chapter-4** | Aromatic Compounds | **59** |

# Chapter-1:

## Nucleophilic Substitution Reactions

## Introduction:

Nucleophilic substitution reactions can be categorized as;

- ➢ **Unimolecular substitution reactions ($S_N1$)**
- ➢ **Bimolecular substitution reactions ($S_N2$)**

## Unimolecular substitution reactions ($S_N1$)

- $S_N1$ is two step reaction and proceeds via carbocation intermediate
- Rate determining step involves only single reactant
- $S_N1$ yield racemic mixture
- In $S_N1$, steric effects are of no significance
- Neighboring group participation in $S_N1$ can be important

## Bimolecular substitution reactions ($S_N2$)

- $S_N2$ is single step reaction, attack of nucleophile and departure of leaving group takes place simultaneously.
- Rate determining step involves two reactants
- $S_N2$ results in inversion of stereochemistry
- In $S_N2$, steric effects are of much significance

**Important Note:**

Generally, tertiary carbons undergo *via* $S_N1$ and primary carbons undergo *via* $S_N2$.

Secondary carbon may undergo via $S_N1$ **or** $S_N2$ depending on other factors like

solvent polarity, nature of nucleophile and leaving group.

# A STEPWISE APPROACH TO SOLVE PROBLEM

**Step-1:** Label all electrophilic and nucleophilic sites present in reactants and products. Also identify acidic and basic sites.

**Step-2:** If more than one electrophilic or nucleophilic sites are present then identify the most reactive site

**Step-3:** Recall all reactions of the most reactive functional groups and decide which is the most suitable while taking reaction conditions in account.

**Step-4:** Work through the mechanism, which lead to the intermediate product.

**Step-4:** Repeat the above four steps

**Step-5** Recognition of structure which is closely related to the final product

**Step-6** Write down the structure of the product

**EXAMPLE-1.1**

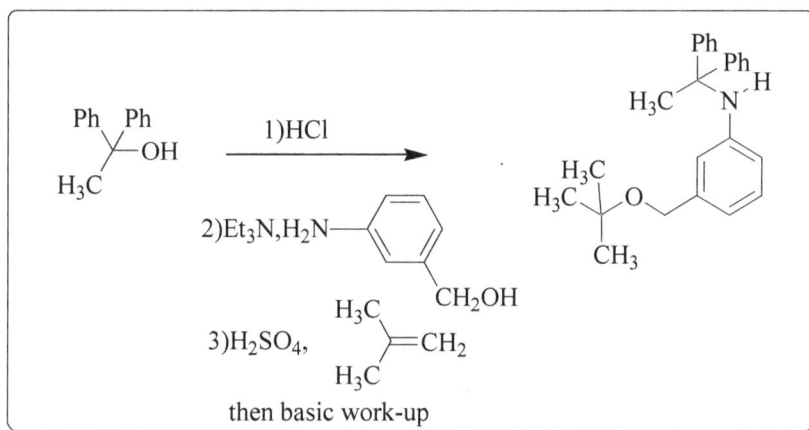

**Important hint:** Substrate possess tertiary carbon so it will undergo *via* S$_N$1

**SOLUTION OF EXAMPLE-1.1:**

**Step-1:** In this example, alcohol is nucleophile and base because oxygen has lone pairs. HCl is strong acid and fully ionized so Cl$^-$, may act as nucleophile.

**Step-2:** Aromatic rings can b protonated but alcohols is the most basic

**Step-3:** Alcohols can be protonated easily using strong acid.

**Step-4:** Leaving group (water) departs apart, leaving behind resonance stabilized carbocation

**Step-5:** Carbocation readily undergoes $S_N1$ reactions and carbocation is entrapped by the most reactive nucleophilic functional group (amine); a series of proton transfer gives the product. Isobutene is protonated to give tertiary carbocation, which is captured by alcohol to give required product in protonated form

**Step-6:** Deprotonation of the resulting ether gives final product

# REACTION MECHANISM OF EXAMPLE-1.1

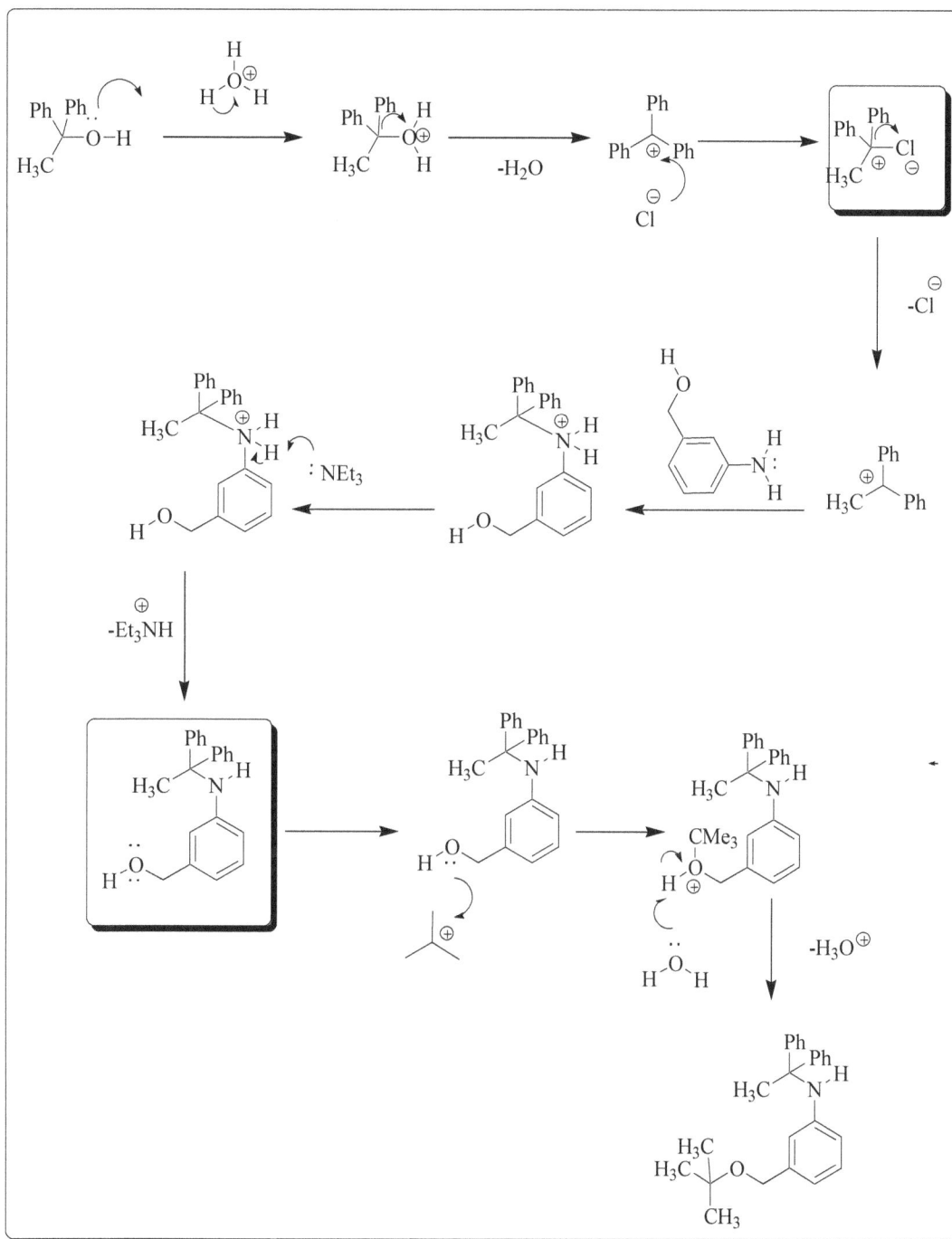

# PRACTICE PROBLEMS

**Please, mention the products in the following reactions**

**Problem-1.1:**

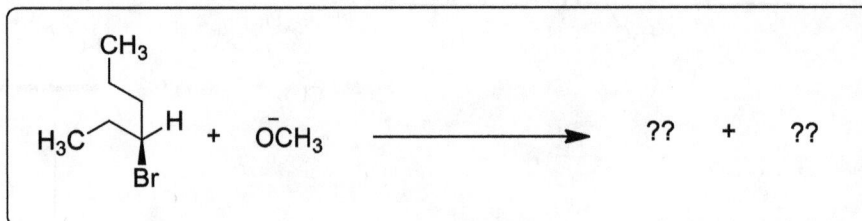

**Problem-1.2:**

**Problem-1.3:**

## Problem-1.4:

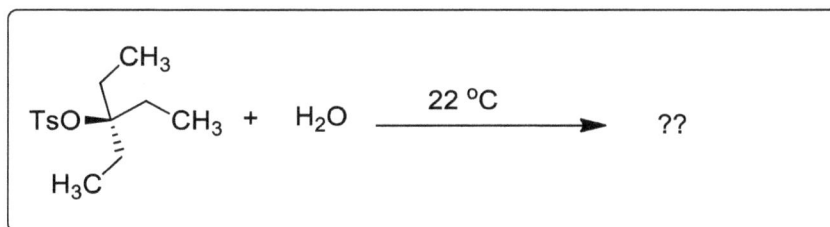

## Problem-1.5:

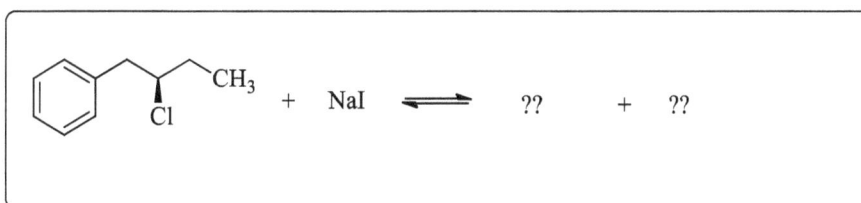

## Problem-1.6:

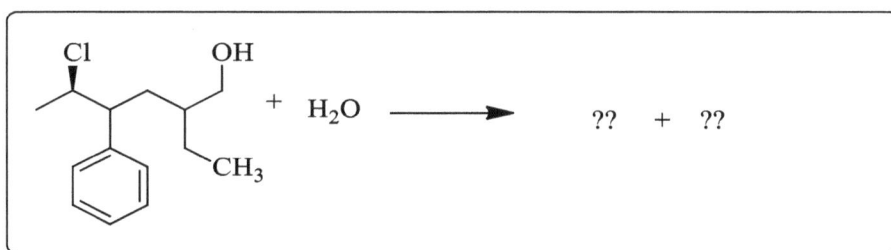

## Problem-1.7:

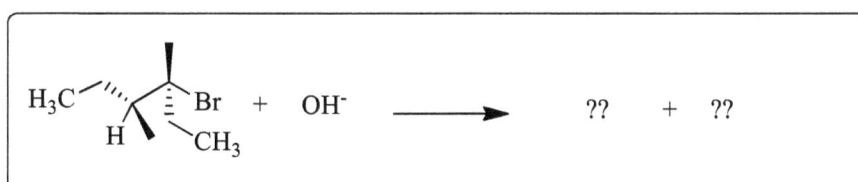

**Problem-1.8:**

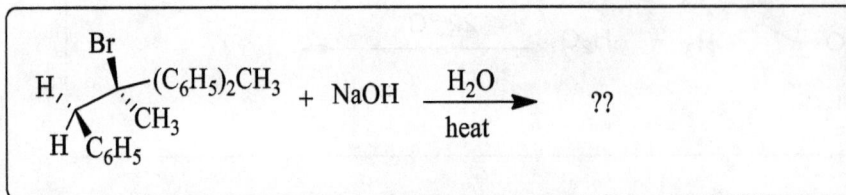

**Problem-1.9:**

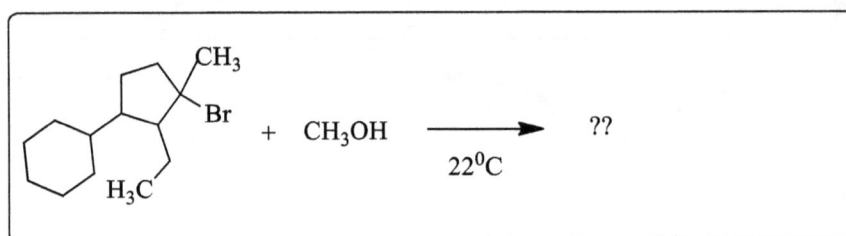

**Problem-1.10:**

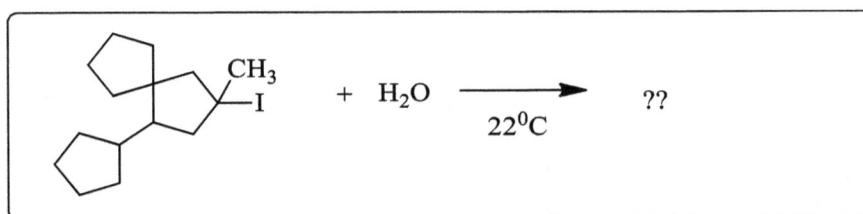

**Problem-1.11:**

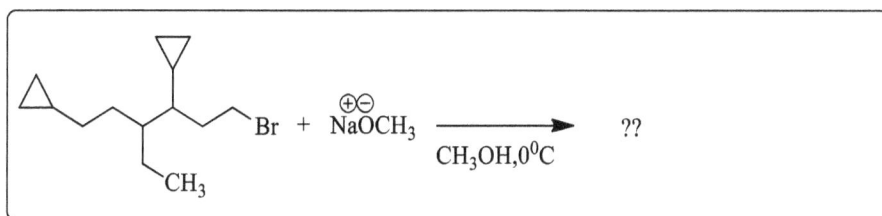

**Problem-1.12:**

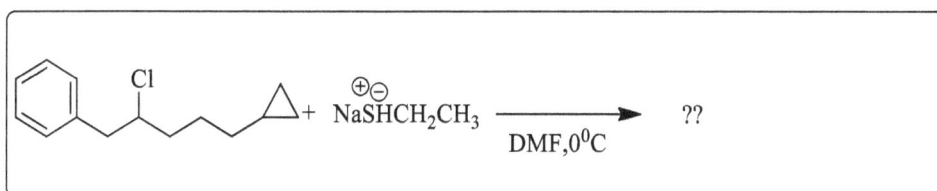

**Problem-1.13:**

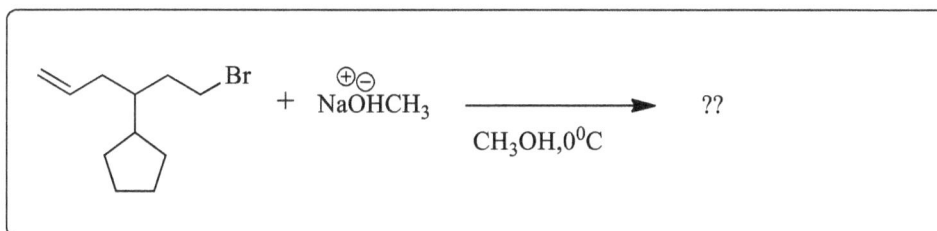

**Problem-1.14:**

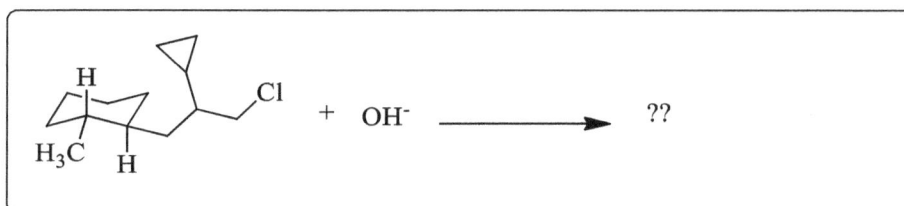

**Problem-1.15:**

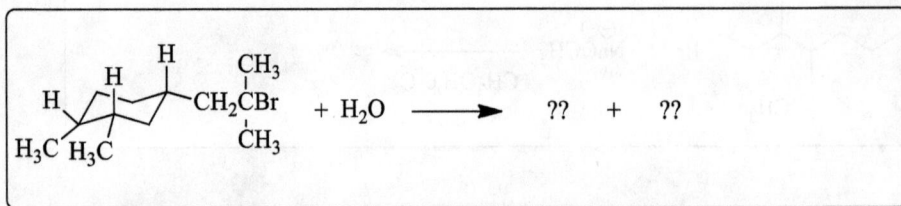

$$\text{(structure)} + H_2O \longrightarrow ?? + ??$$

**Problem-1.16:**

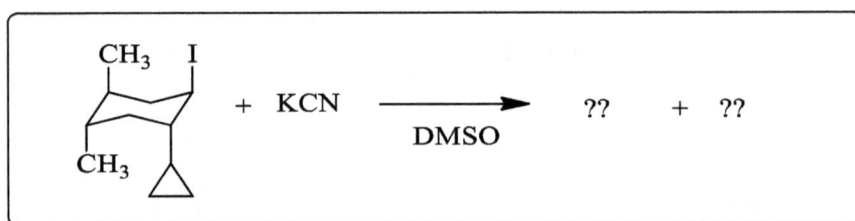

$$\text{(structure)} + KCN \xrightarrow{\text{DMSO}} ?? + ??$$

# SOLUTION OF PRACTICE PROBLEMS

**Solution-1.1:**

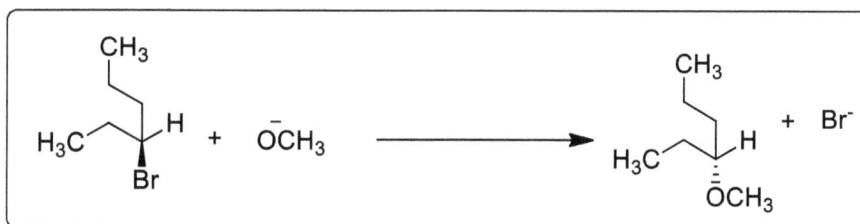

**Solution-1.2:**

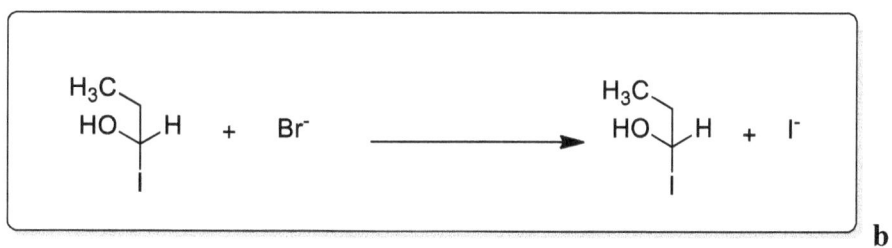

b

**Solution-1.3:**

**Solution-1.4:**

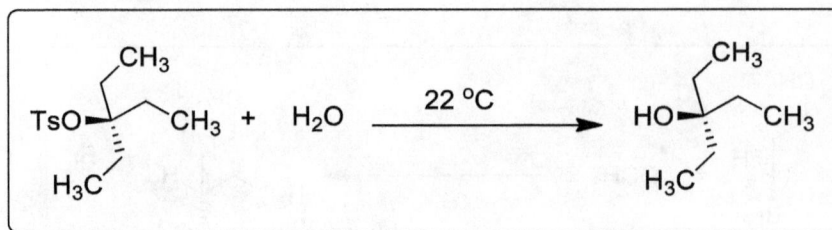

**Solution-1.5:**

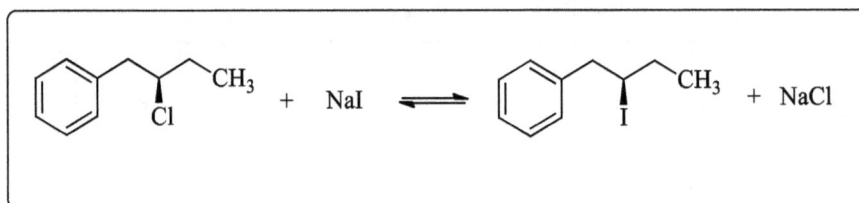

**Solution-1.6:**

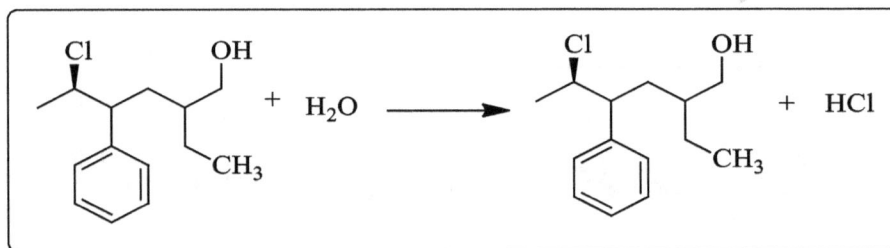

**Solution-1.7:**

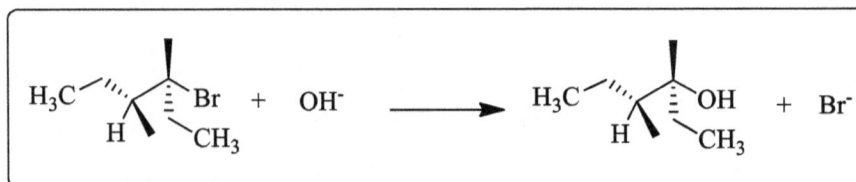

**Solution-1.8:**

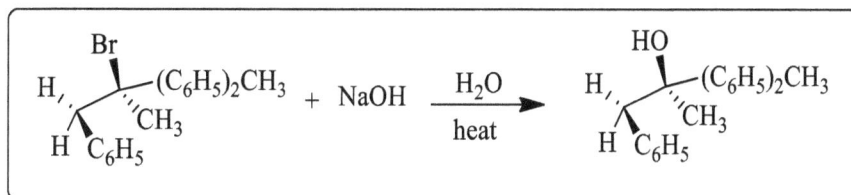

**Solution-1.9:**

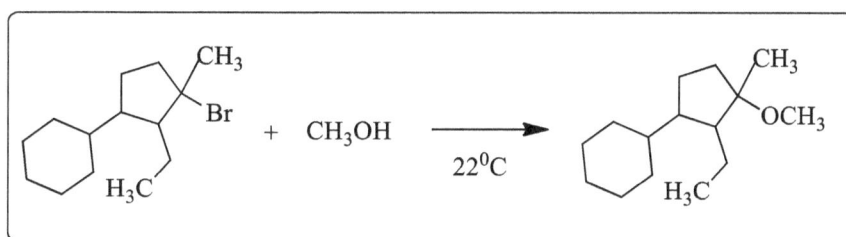

**Solution-1.10:**

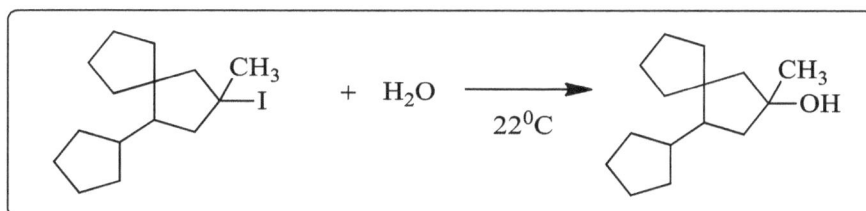

**Solution-1.11:**

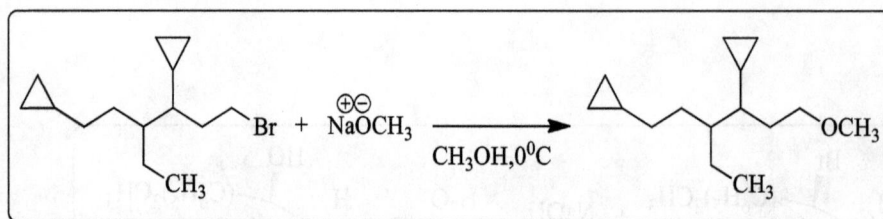

**Solution-1.12:**

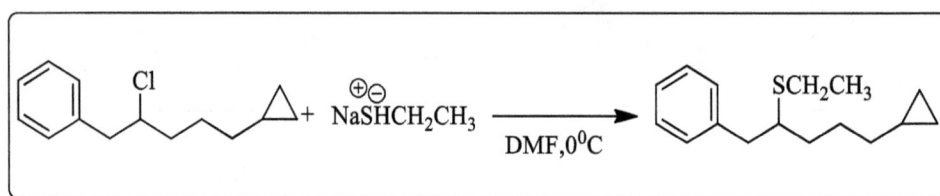

**Solution-1.13**

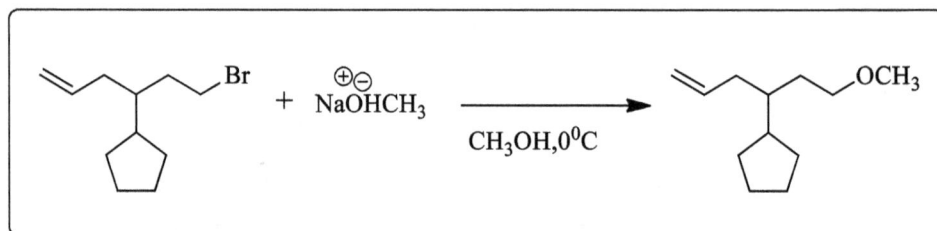

**Solution-1.14:**

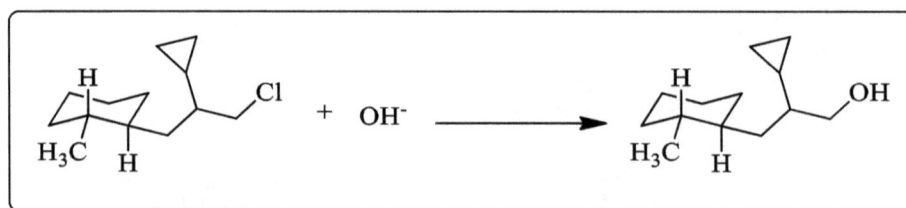

15

**Solution-1.15:**

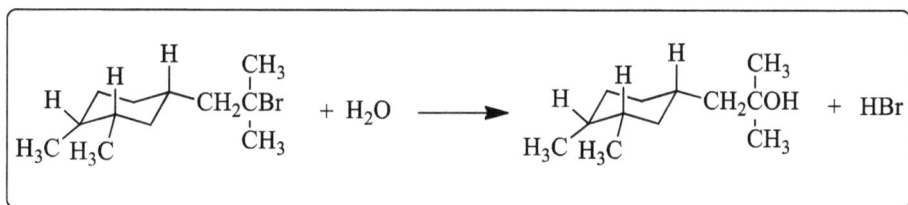

**Solution-1.16:**

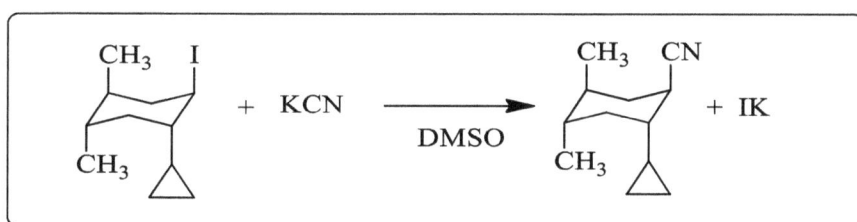

# Chapter-2:

## Nucleophilic addition to carbonyl groups

**Introduction:**

All compounds having carbonyl groups may undergo nucleophilc addition reaction. Compounds having carbonyl group can be categorized as;

- **Aldehydes (R-CO-H**
- **Ketones (R-CO-R)**
- **Acid (R-CO-OH**
- **Esters (R-CO-OR)**
- **Amide (R-CO-NH2)**
- **Acid halide (R-CO-X)**
- **Acid anhydride (R-CO-O-CO-R)**

**Aldehydes & Ketones**

Addition of nucleophile to carbonyl compounds can be irreversible and reversible.

**Irreversible addition reaction;**

1)-Reduction of carbonyl group

2)-Cabanion addition to carbonyl group

**Reversible addition reaction;**

1)-Addition (addition of cyanide, bisulfite etc)

2)-Addition-substitution (Important for group VI nucleophiles, i.e.; oxygen and sulfur leading to respective hemiacetal, acetal and thioacetals)

3)-Addition-elimination (Important for group V nucleophiles, i.e.; ammonia, hydrazine to give imine and hydrazine respectively)

## Acids, Esters, Amides & Anhydrides

All these carbonyl containing compounds have a leaving group so every nucleophilc adition to carbonyl compound accompanies elimination. The net reaction is addition-elimination through a tetrahedral intermediate. They regains carbonyl carbon after departure of leaving group.

For nucleophilic addition to carbonyl group, the reactivity order is;

**R-CO-Cl > R-CO-O-COR > R-CO-H > R-CO-R > R-CO-OR > R-CO-NR$_2$**

# A STEPWISE APPROACH TO SOLVE PROBLEM

**Step-1:**    Label all electrophilic and nucleophilic sites present in reactants and products. Also identify acidic and basic sites.

**Step-2:**    If more than one electrophilic or nucleophilic sites are present then identify the most reactive site

**Step-3:**    Recall all reactions of the most reactive functional groups and decide which is the most suitable while taking reaction conditions in account.

**Step-4:**    Work through the mechanism, which lead to the intermediate product.

**Step-4:**    Repeat the above four steps

**Step-5**    Recognition of structure which is closely related to the final product

**Step-6**    Write down the structure of the product

**Example-2.1:**

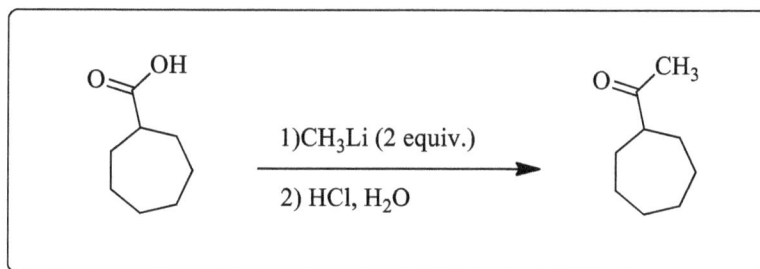

**Important Hint:** CH₃Li is a strong base and also act as good nucleophile

**SOLUTION OF EXAMPLE-2.1:**

**Step-1:** Hydrogen of carboxylic group is acidic in nature and methyl litjium is a very strong base and nucleophile.

**Step-2:** In this example, carboxyl group is the only reactive site.

**Step-3:** Always acid-base proton exchange is the most convenient reaction so it happens earlier than Nucleophilic attack if there is possibility of both.

**Step-4:** nucleophile adds to the carbonyl carbon and gives a tetrahedral intermediate that forms ketal after acid treatment.

**Step-4:** Ketal undergoes dehydration in the presence of acid.

**Step-5** Deprotonation yield the required product

**Step-6** Ketone is the product

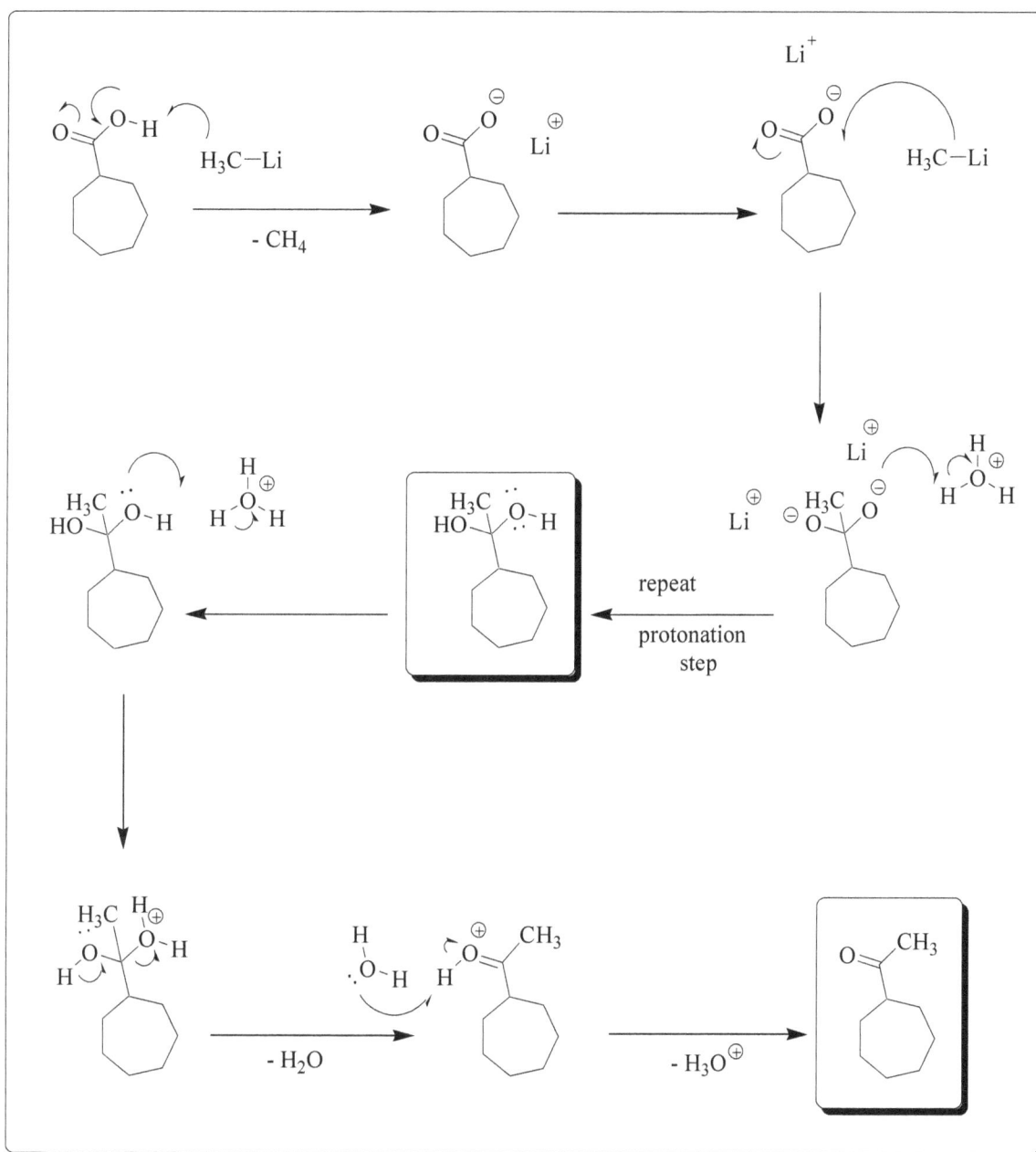

# PRACTICE PROBLEMS

**Problem-2.1:**

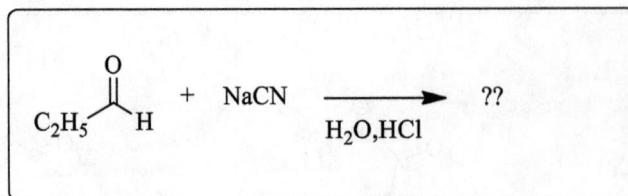

**Problem-2.2:**

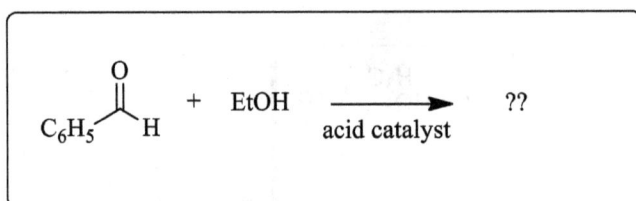

**Problem-2.3:**

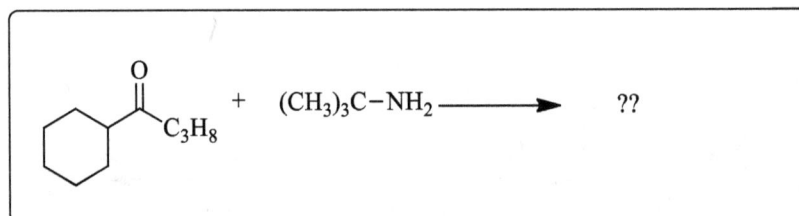

**Problem-2.4:**

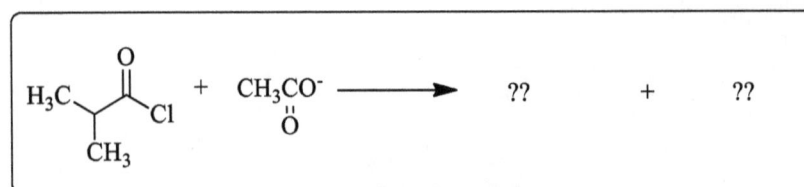

**Problem-2.5:**

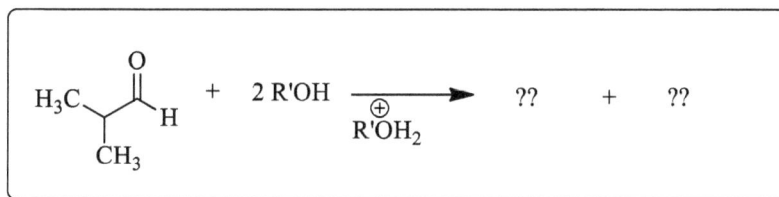

**Problem-2.6:**

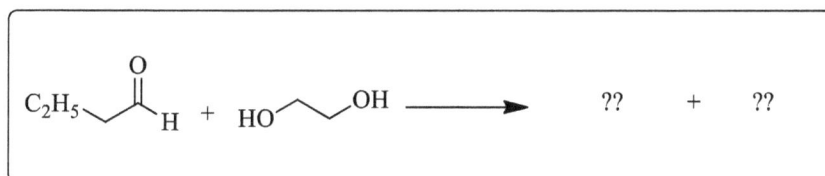

**Problem-2.7:**

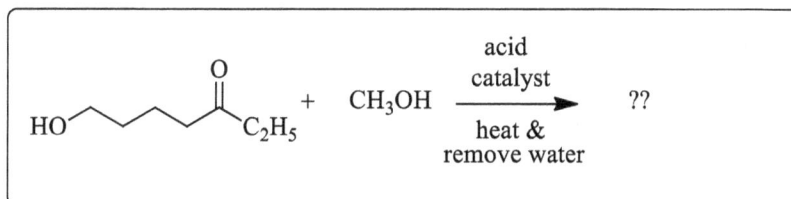

**Problem-2.8:**

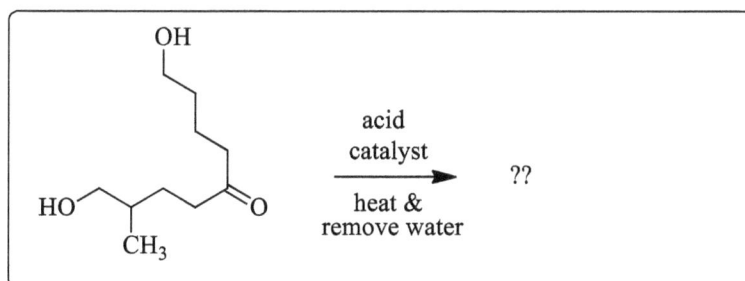

**Problem-2.9:**

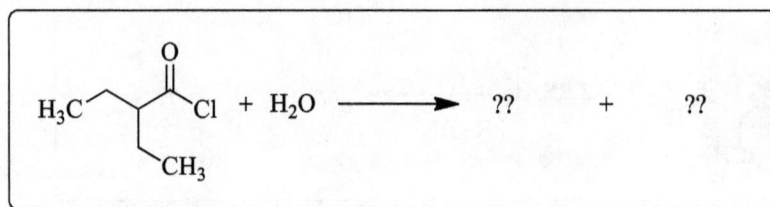

**Problem-2.10:**

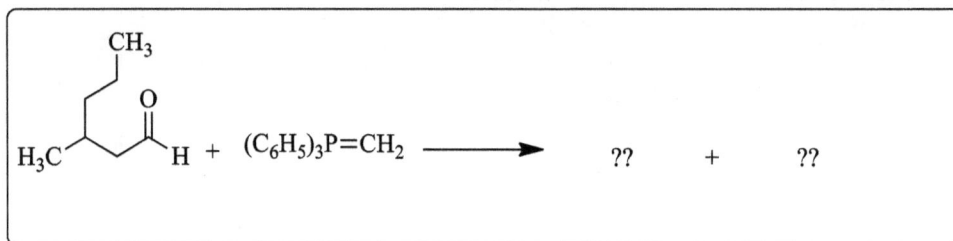

**Problem-2.11:**

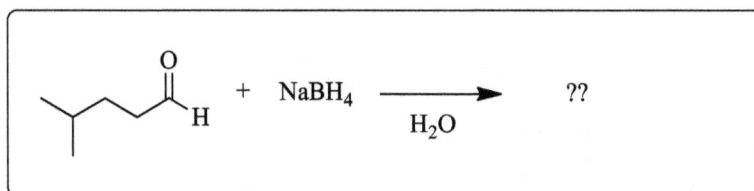

**Problem-2.12:**

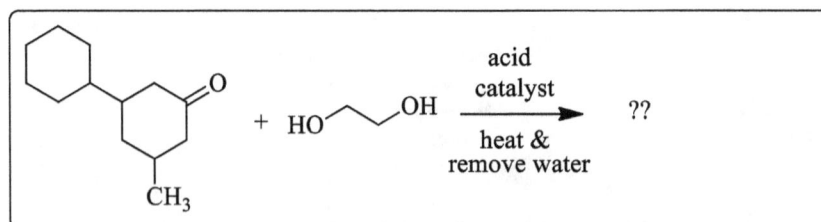

## Problem-2.13:

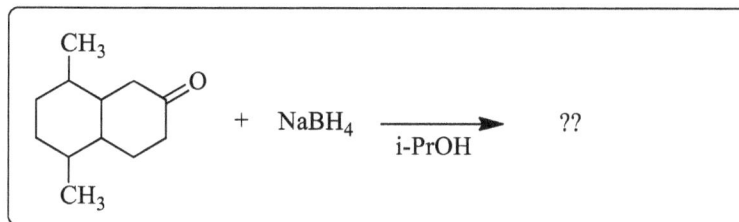

## Problem-2.14:

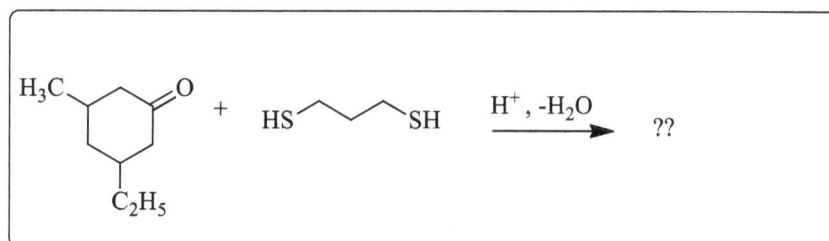

## Problem-2.15:

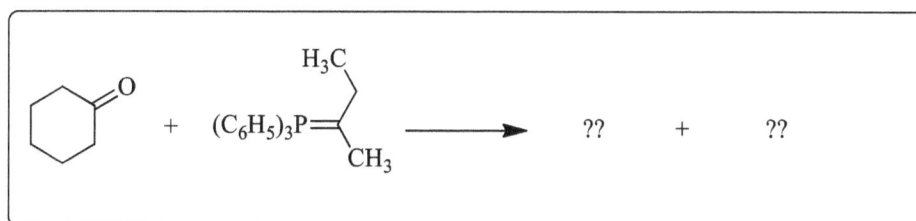

**Problem-2.16:**

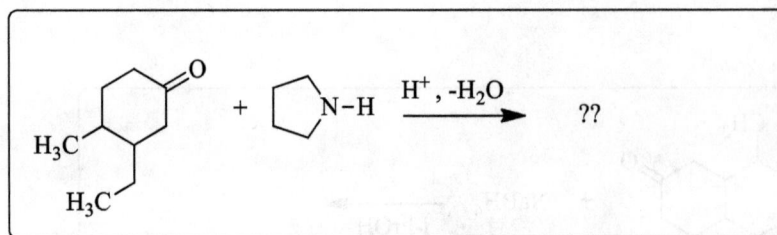

**Problem-2.17:**

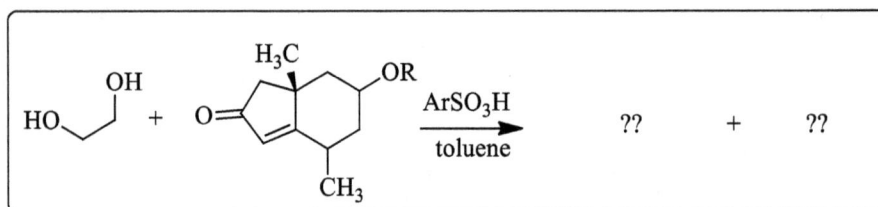

**Problem-2.18:**

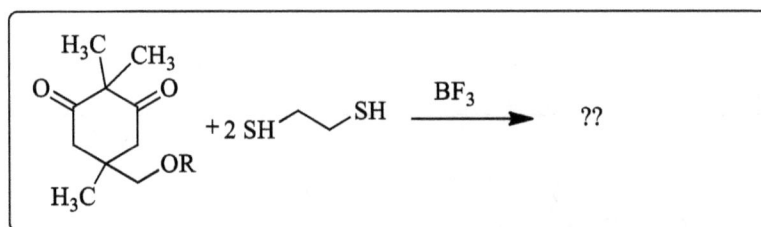

**Problem-2.19:**

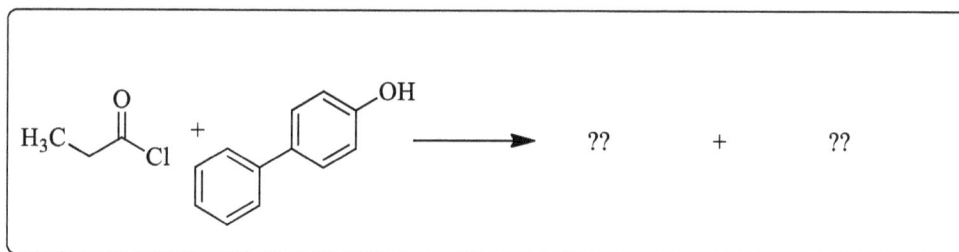

**Problem-2.20:**

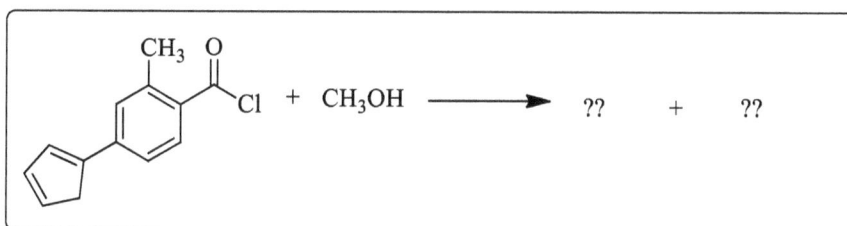

**Problem-2.21:**

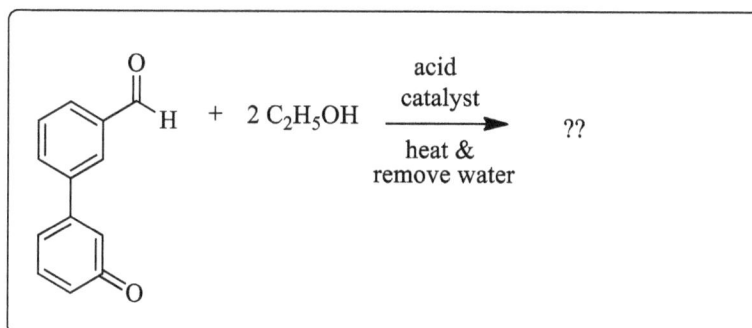

**Problem-2.22:**

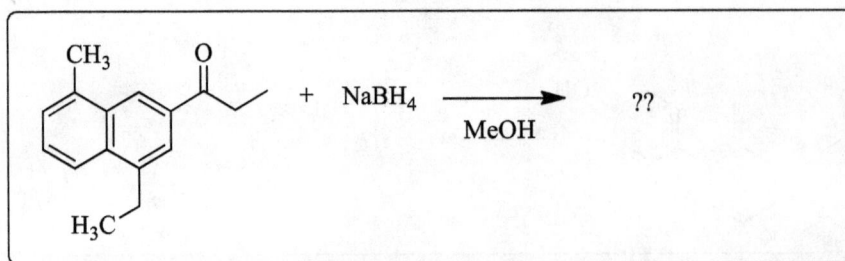

**Problem-2.23:**

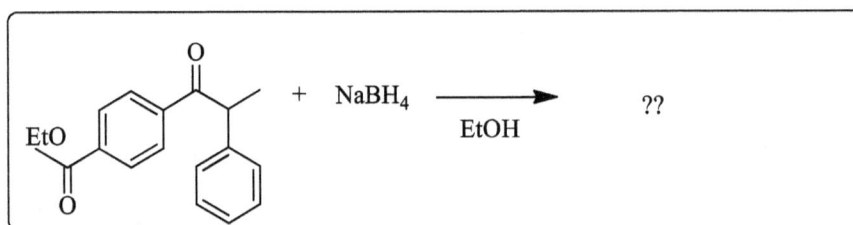

**Problem-2.24:**

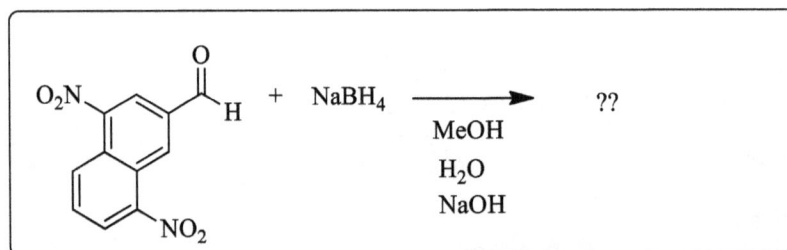

## Problem-2.25:

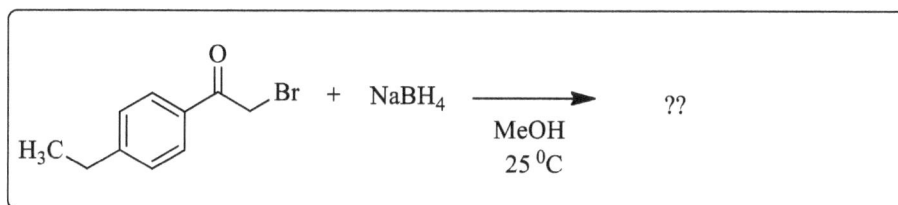

## Problem-2.26:

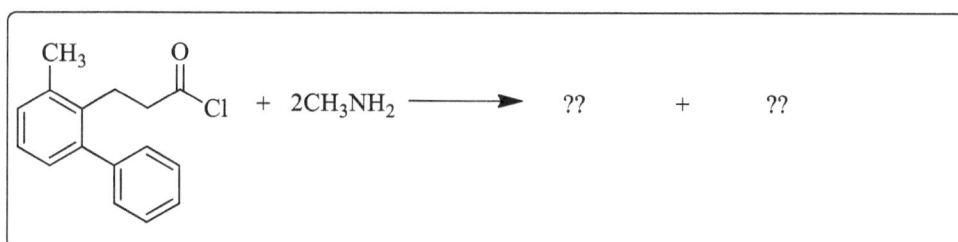

# SOLUTION OF PRACTICE PROBLEMS

**Solution-2.1:**

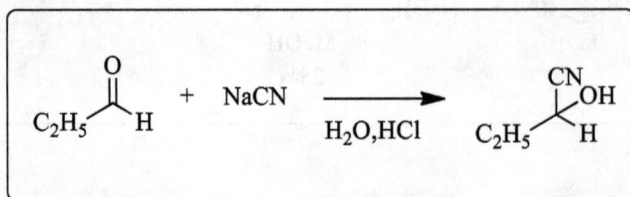

**Solution-2.2:**

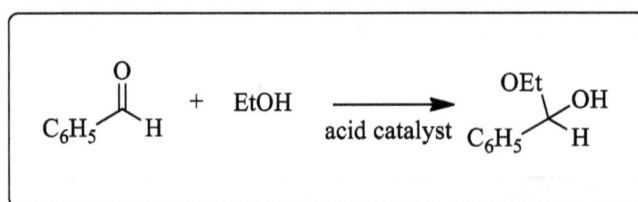

**Solution-2.3:**

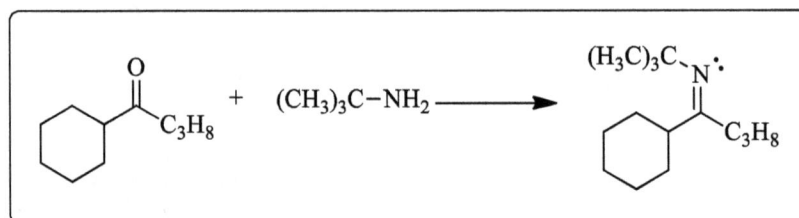

**Solution-2.4:**

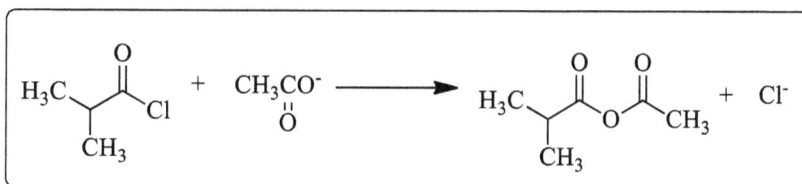

**Solution-2.5:**

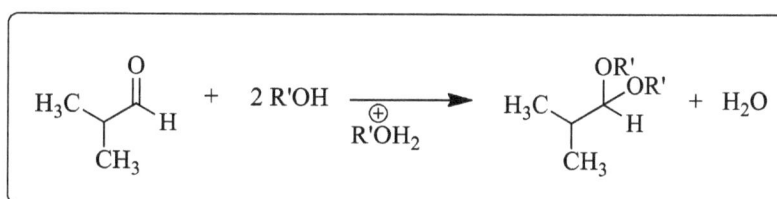

**Solution-2.6:**

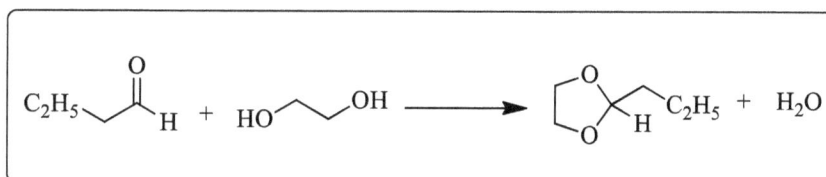

**Solution-2.7:**

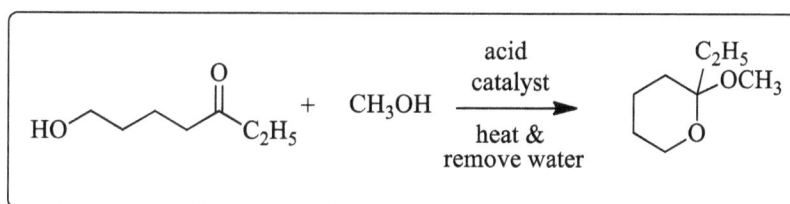

**Solution-2.8:**

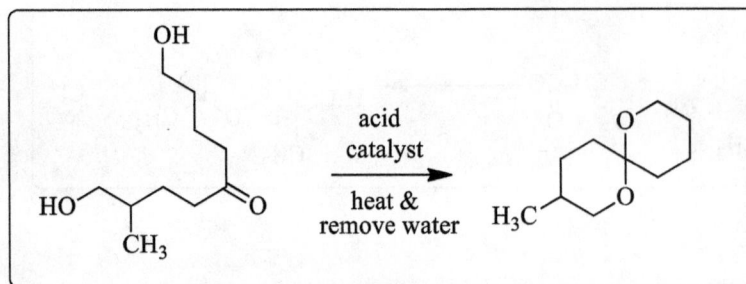

**Solution-2.9:**

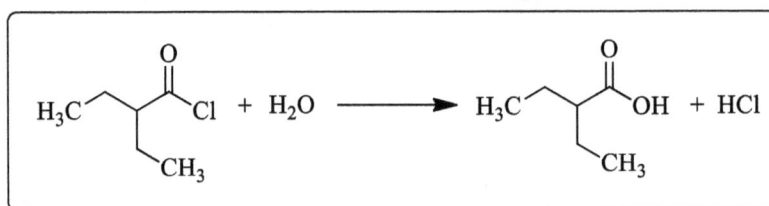

**Solution-2.10:**

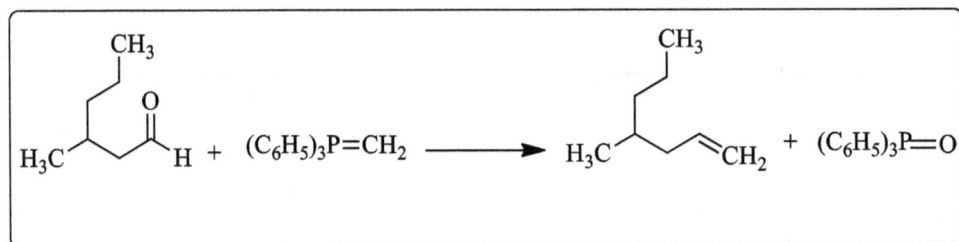

**Solution-2.11:**

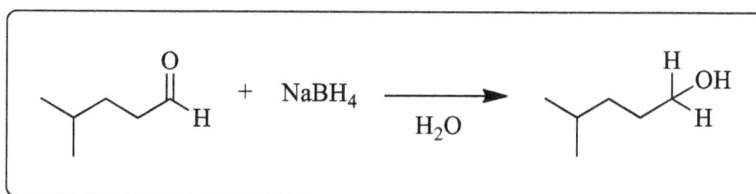

**Solution-2.12:**

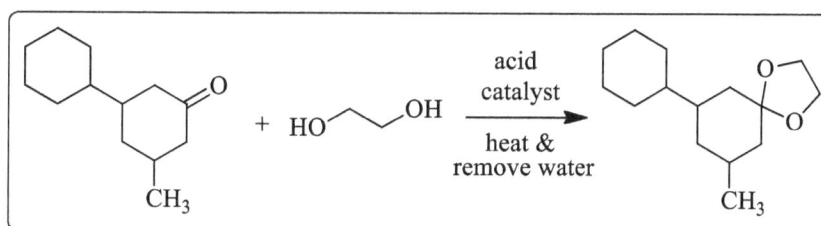

**Solution-2.13:**

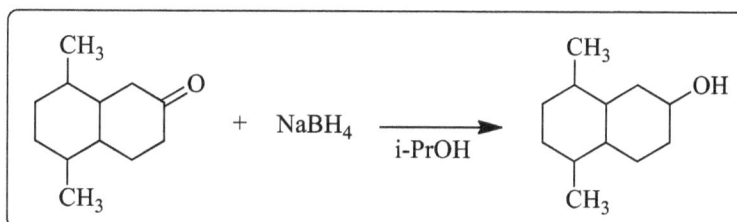

**Solution-2.14:**

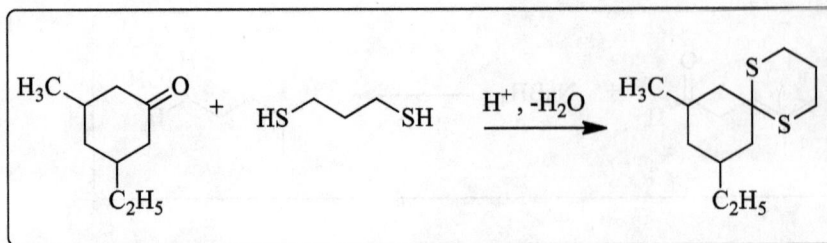

**Solution-2.15:**

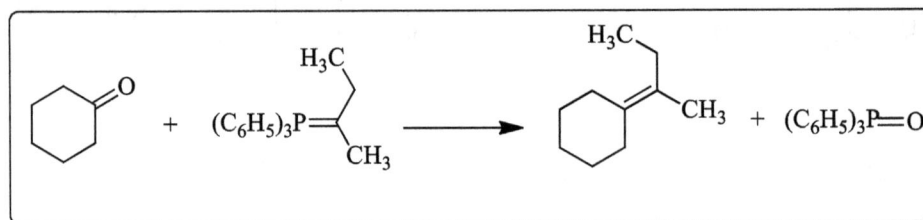

**Solution-2.16:**

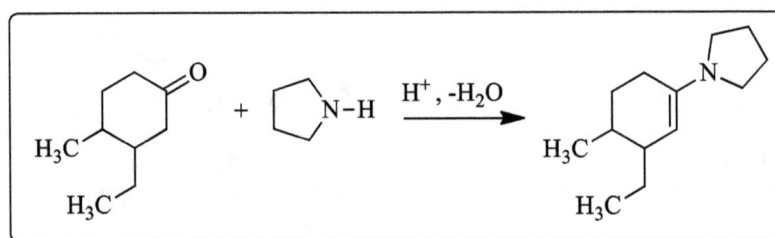

**Solution-2.17:**

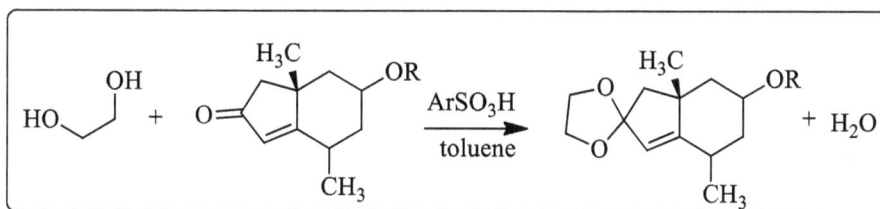

**Solution-2.18:**

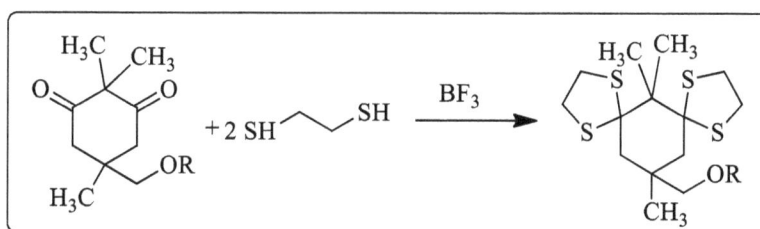

**Solution-2.19:**

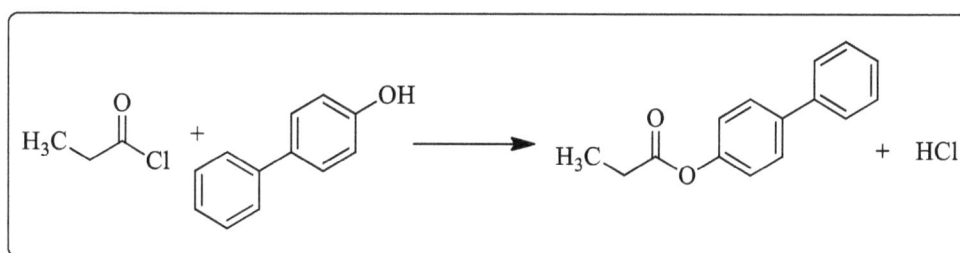

**Solution-2.20:**

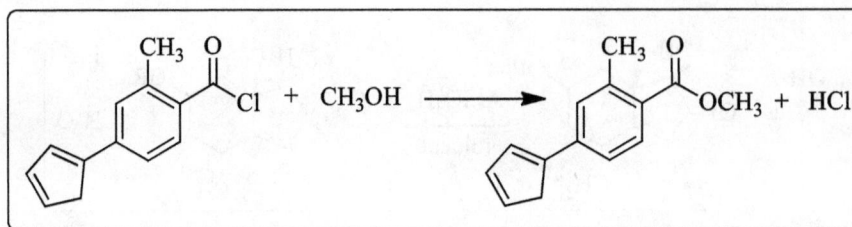

**Solution-2.21:**

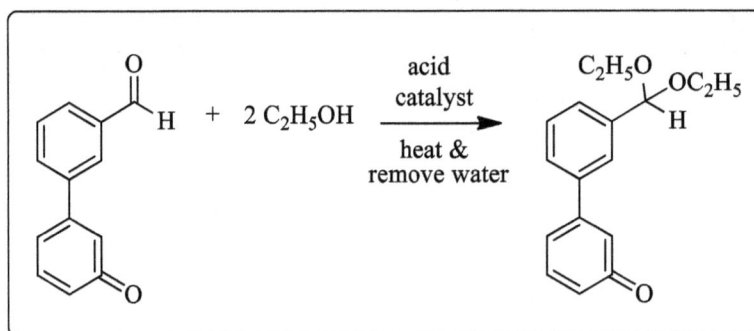

**Solution-2.22**

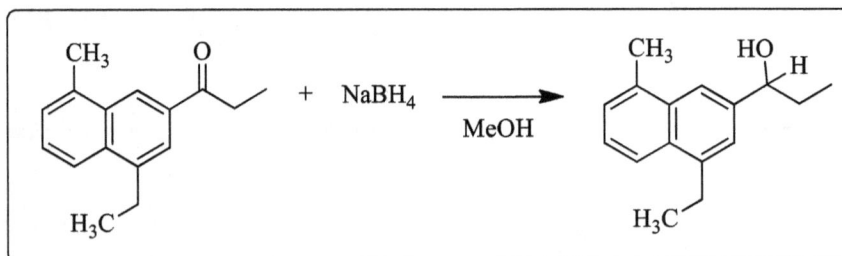

**Solution-2.23:**

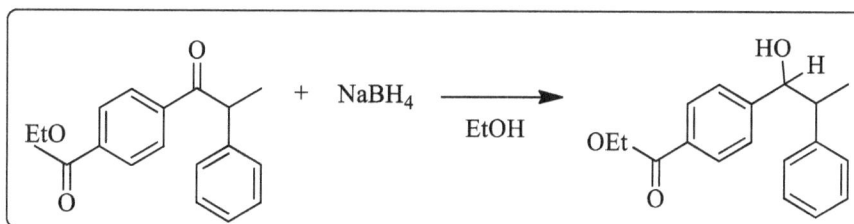

**Solution-2.24:**

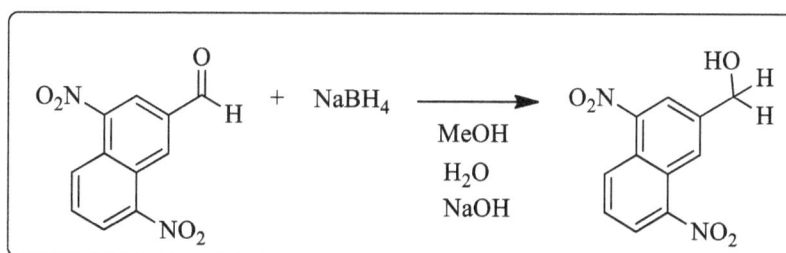

**Solution-2.25:**

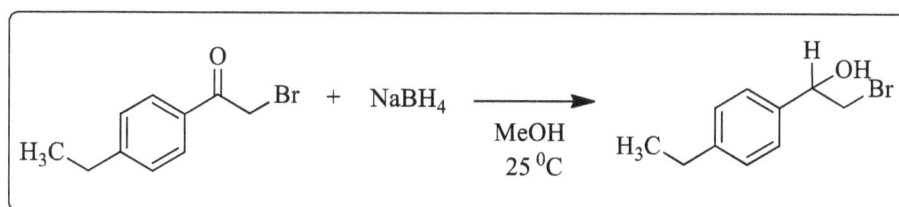

**Solution-2.26:**

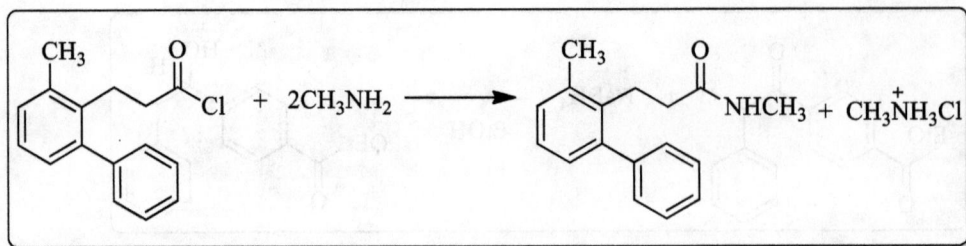

www.ingramcontent.com/pod-product-compliance
Lightning Source LLC
Chambersburg PA
CBHW082114210326
41599CB00033B/6694